TRIANGULATION
FROM A KNOWN POINT

MW00569546

RUTH DANON

North Star Line Poetry Series
Edited by Jay Parini

NORTH STAR LINE/NEW YORK

First Blue Moon Edition 1990
First Printing 1990

ISBN 0-929654-97-8 (Hardcover)
ISBN 0-929654-93-5 (Paperbound)

Manufactured in the United States of America.

A North Star Line Book
Published by Blue Moon Books, Inc.
61 Fourth Avenue
New York, New York 10003

PREFACE

This first book of poems by Ruth Danon is singularly fine. Each poem gathers tightly about an image, or a sequence of linked images; these images, cut from the broadcloth of Danon's unusually rich diction, radiate, double back on themselves, and reach out once again. The effect is like a stone tossed into an icy pond; it disappears, but the rings around the struck center multiply, lapping the far shores of the reader's consciousness long after the book is put away.

Though she has thus far published only in literary magazines, Ruth Danon is a mature poet already, and Triangulation from a Known Point bears no resemblance to most first books in that her models have been fully absorbed, transmogrified, made into something entirely her own. Nevertheless, literature is a tissue of allusion and influence, and all poets come from somewhere. The obvious debt in Danon's work is to the tradition that runs from Baudelaire and the French Symbolist poets through Wallace Stevens, Georg Trakl, Pablo Neruda and, especially, John Ashbery. This is to say that her poems proceed, not in any linear fashion, but through an accretion of imagery. The poems resist paraphrase, their full meaning gathering in the margins, underneath the playful surface.

There is a also a sense of perpetual surprise here, with Danon juxtaposing words and phrases in uncanny ways, making huge leaps of association. The syntax is equally surprising, the lines snaking through unexpected tracts of feeling and thought, as in "Keeping Track," which begins:

> Distance to distance. The sound of
> thunder, a train, recedes to the west.
> Your hand running the length of my arm.
> There is no beginning to an action.
> You touch me, the sheet, your own
> shoulders. We make love, seamless,

successful as the illusion of cinema.
"We have always been here," you tell me.
There we were. I am in a cafe
In New York in early spring.

On and on, the unwitting reader steps through the bright
crust of Danon's poetic surface and falls headlong, like Alice
into her Wonderland, into an abyss of experience that mingles
joy and pain in almost equal measures.

In the patchwork quilt of Danon's world, one theme
emerges like a watermark: the search for bearings, accompanied by a profusion of geological, geographical, cartographical references. The poet is "triangulating" from a known
point, lifting the sextant of her imagination to the stars. She
is trying to discover by any means available the longitude
and latitude of her immediate position. Poetry is, for Danon,
a form of dead reckoning, a coming to terms with the crude
realities, the parameters, of her existence. Danon begins, as
she must, with what she already knows. And what she knows
is this, that "nothing sufficient has yet been said about stone:
/ its weight, its mass, its edge, / its impact on bone."

Triangulation from a Known Point is a devilishly good
book: direct, inventive, wild, and engrossing. It is keyed and
pitched in a way all its own, a poetry that earns its keep at
every turn and makes one eager for another book by this
inventive and accomplished poet.

JAY PARINI

FOR VERA POLANYI-DANON
(1909–1987)

Acknowledgment is made to the editors of the following magazines in which many of these poems first appeared: *Another Chicago Magazine: Bomb; The Boston Review; The New England Review; The Gettysburg Review; The Paris Review; Pink; Tendril*.

I am grateful to the Ragdale Foundation and to the Corporation of Yaddo for giving me the time and the peace of mind to work. Many of the poems in this book were written during residencies at these remarkable places.

Table of Contents

Acknowledgments

Table of Contents

SENSIBLE SHOE

VOCATION

The hardest part is the way the knot at the base of the neck tightens.
There's a whole body here to hold up the head but this time
you are tired and you want to lie down. Never would a gesture
mean so much.

The change is subtler than you expect. You were hoping for
a circus, a pony-ride, the high trapeze. You wanted a presence,
the sudden rush of arrival. The hardest part comes now,
this discovery of needs more radical than food and shelter.
How to say it? There's hell to pay this time and you know
you'll take anything that's offered.

Lately you've had to develop new uses for your college education.
Perhaps you had the wrong major after all, should have stayed in botany,
scooping algae out of the river, scraping moss from bark, mastering
the small details of other life forms.

There were so many uses for geological science. There was such
possibility, you thought, so many secrets. It would have been nice
to read the weather with some intelligence, to know a mackerel sky
when you see one.

      I know this much:
Nothing sufficient has yet been said about stone:
its weight, its mass, its edge,
its impact on bone.

SATURDAY NIGHT IN THE NEW WORLD

Someone looks out a window, swallows
a map. Someone forgets the orders.
Persistent as snow falling in Gdansk,
Molo rolls over into a new life.
Snowplows have not yet been invented
and there is cold water in every well.
All night things happen without charity:
the army squats low like a rat in a hole
and girls in leg warmers walk
from the north corner to the south.
Someone imagines that geology
is an exact science.
Molo is dreaming of a good night's sleep
but a light is burning on the fifth floor
and shines in her eyes.
Molo is lacking a cat, among other things.
She remembers a saxophone
and says the word several times.
Certain skills are required
that no one has ever learned.
People have discovered
interesting new sexual practices
and many books have been written
about the way chains dominate the imagination.
Molo wants out. She wants the flat edge
of her childhood: a day in the garden,
the path to the stake that almost
put her eye out when she reached for the fruit.
She wants the daisies that pulsed in a field
and were too far to reach. She wakes
to her uncle shaking her from sleep,
leading her into the cold, up a plowed

hillside to see, barely there,
those rare colors in the night sky.
Aurora borealis, nothing lasts.
Molo has been thinking
for a long time.

ALASKA

In summer she cuts her hair
to learn the shape of her skull.
Someone who is not her lover
teaches her to drink whiskey
straight. She gulps it down
in paper cups. It burns her throat.
At night she dreams of Anchorage,
everywhere Alaska and clean.
She rides the slow-moving trains
pushing through snow, across Canada.
It's the big silence she wants,
the kind that can't be found
when things stay light for long.
Her lover says "In Anchorage
it rains. It's never cold. Everywhere it's soot."
Her dreams change. Now she flies
north of Fairbanks to Eagle.
Here the sky stays dark all day.
Soft fur grows on her feet and hands.
Patiently she follows the brown
bears as they lumber through snow.
They lead her to the edge of frozen
water. She counts the fish in the ice.

6

SENSIBLE SHOE TAKES A WALK

Something to do with architecture,
she thinks, or freedom. Something
grand like that or like a little dog
dreaming itself a cloud. The way
fish dream of bread, warm loaves
on the water. There was another life
and you know it. Casting stones.
The way hope sinks. The way it is.
Only this fat spring, violent as a
slug, and the rippled air of the moon.

frames us
frameless

biblical

looking up from below

2 worlds

world of forms &

the dream world.

Dream world: free?

the contradiction the contradiction in her metaphors

fish don't dream of bread because they have no sense of forms.

Creating illusions around ourselves, being dogs to a formless world.

parallels (Sensible Shoe takes a walk and Half Way

we are only senses, subjects persons, with need — not objects

7

SENSIBLE SHOE IN THE SITUATION ROOM

Not everyone is having a nice day.
Behind my eyes the first
articulate scratch on the surface
of this sweet globe. I wanted
to draw you a map of the world
before it was made. I would
spare you this terrible early dark.
A desire to tell the truth
overwhelms any simple under-
taking of geography.
Neither the beating to death
of monkeys on the high plains
of Africa nor the blinding
of rabbits in white rooms
is evidence enough. The earth
is a clenched fist at the end
of an outstretched arm. The innocuous
orange spins off. Forget this air,
this gas, this molten core.
The dumb stars blink.
This is my job:
to lay it out
without distortion.

SENSIBLE SHOE IN BASIC TRAINING

Fifteen parts of the house are hiding
but so what? Aren't you glad of adventure,
all the geological or archaeological or
otherwise investigations of what others
would have thrown away? For years now
I have been taking myself to the cleaners.
The cat's got my tongue, desire trapped
in the round green eye of willfulness.
Every return to unfamiliar ground has
its price. Love's what remains after
the sands have shifted; how it feels
to come across shards of glass or painted
bits of clay or bead. Information can be
bartered for evidence. A bird in air
catches the light, breath harboring sound.
This is a conspiracy of magnificent proportions.
I need a new moon in my window. I need
a better heart. The cat's at bay.
She faces the glass singular as a pear
on a blue plate. A walk outside won't
clear up anything. Cold takes its course.
I'm all burned up. Discipline's the latest
trend. I don't get it. I was just
looking for the door.

AFTERIMAGE

glass is the cause of her transformation into a doll — constant need of reflection — vanity

Doll implying lifeless a ghost — artifical

If not for glass
the doll propped up
in the window of the top
floor of the Hotel Amsterdam

Amsterdam like a hotel — resort rooms

would flop belly first into the wet street.
For three sullen days
I have watched those brave fat arms

she's the doll about to fall is she on the wet street.

stretched out like a diver.
And each night of the sullen three,
alone in this pitched room, I dream of him.

key line — makes the poem

I swear, three years dull as the flu,
there aren't enough facts in the Book of Fishes
to account for the absence of air in my lungs.

can't breathe

When I wake into the debris of illness *residue*
it's the way I felt when I saw I'd been robbed:
all bright objects: lamp, radio, clock, camera, in corona, *isolated*
still there, the long bubble of a voice through water
or the orange staying on after blue.

3 days reminds her of three years.

*lamp, radio, clock
light, music, time — all seemed "bright"*

Alone in Amsterdam wishing she was with "him."

all the objects of her past have been robbed, not by him, but by her illness of not being able to get past that void — sickness — depression

or maybe she closes wants to be with him just to find freedom from loneliness.

10

SENSIBLE SHOE ABANDONS THE PATH TO WISDOM

Silence and the bright activity of hands.
The man washing oranges
outside the bird of paradise shop on a cloudy day.
There will not be rain. There were predictions.
This is the edge of displacement. There is the need
to assess the damage, not yet determined.

The damage hasn't yet been determined. Weighing
grapes, hope for a pleasant dinner; think
of your lost talents. Books are full of facts;
the earth at any one moment can have but one
configuration. Errors of similar magnitude
have been made in the past.

SENSIBLE SHOE DEVELOPING SIXTH SENSE

Hanging on to implications
gets me down, waiting
night after night
trapped in this grid
for that singular occurrence.
My feet are sore
after so much
walking around in a circle.
You don't have to be lost
to take note of latitudes.
The presence of Shetland
ponies is no indication
of place or anything else.
There are bright dishes
in space, saucers flying around.
Not on a lost island,
not in a sheltered cove,
I can feel the most recent
shift of land, bullet from gun.
Practically speaking, everyone
deviates. I concentrate
on the visible migration
of birds; this has been
of assistance.
      Compass, sextant, radar, sonar: the words
seduced me, were
the wrong tools. Predominant
activity of the bad hand
persists in spite of training.

MEMORANDUM

The fact of a wildlife preserve
makes only a marginal difference
in the daily lives
of most animals.

) facts are
hasty fallacies
that smear natural
beauty

HALF WAY

Sensible shoe is like having the imagination inserted into a walking body.

objects vs senses.

Suppose an ache, an endurance,
something important long remembered:
Sensible Shoe on the trail of it.
Such upheaval determines
domestic life as we know it. Everything is
in danger, the carnations stiff
in the vase, the cat stretched out
flat on the rug like a map of a foreign country.
Nothing was recovered exactly. The buses are
angry, the trains are angry, and I
am only one woman confronting
the unfamiliar lightness of my own hands.
In this newborn dark I am reading
braille for the first time. I have been
searching for a plum tree that has bloomed
on the side of a mountain in China
for ten thousand years.
The possibilities are that old.
I want my love of the world
sweet as the shadow of that mountain.
The city lays claim; objects and persons
clustered together demand protection.
Consideration of the infrastructure
comes later. This is just beginning.

not punishing stranger

sensing

every human had plenty a sensible perhaps potential.

P 7 X

no time to very hunger passion

not knowing what it is — love

preservation

to look back on the shadowy figures and say it is love

Sensible shoe takes a walk (p 7)

totally nietzsche

14

THE DISCOVERY OF KAFKA'S LIBRARY

Anything for the moment taking me
away from the facts. In the subway
I was reading the paper and not thinking
about the faultline or the rumbling
continents getting ready to move.
Now things will not be the same
at the U.S. Geological Survey. The
maps will be redone; there's one less grave.
Maybe he was a mouselike man or like
a little woollen rat, fond of dark places,
writing letters, glancing at his watch
each day at the same time. An office,
the secret thrill of a secret life.
There was the minute path of his own disease,
fatal and dull as the nineteenth century.
Some news is like adjusting to light,
a couple of blinks before you know
you've been here a thousand times
or never before.

[handwritten annotations: subway tunnel / continents faultlines / critical & / sense / Nietzsche / kafka / philosophy / idea vs. / action or / object / idea of history / maps are words / over continents. / a rendition of / reality.]

THINKS ABOUT SUCCESS

Closes in on the outside chance.
Comes up the stairs. A sigh, ice,
heaves on the pond. The memory,
the key, the turn, the slightest
hesitation. The key, the lock,
the turn to anger. I'm not
begging any more. Little life
shrunk to the size of a room.
Large questions pose
at the window. Will I
get to the beach? Can a cab
take me there?

THINKS ABOUT NUCLEAR DISASTER

I had been thinking about
an ice cream cone. I can't
think in larger terms.
I'd like to get over things, stop
loving the past like some damn
European. I'd bounce back
like an American, build a new city.
Turning the corner, on Monday night,
there were suddenly elephants,
a chain of trunk and tail.
Facing them everyone stopped:
people got out of cabs.
Lumbered stillness and age across 34th
Street. Imagine elephants
doing that.

SHE TELLS THE TRUTH

Invention is not enough.
There are facts
and you need to know them.
In 50 million years
Los Angeles will crowd
into the Aleutian Islands
and chill slowly in the Arctic night.
The Red Sea will dwarf the Mediterranean;
the timing will be better.
There will be no new oceans
and no new continents
but Oaxaca, where you and I
have never been together,
will sink gently into the water.
By then we will have known
each other through
millions of lifetimes. Maybe we'd be wiser
and we'd know what to do.
The equator is where it is
because that position is half way
between the poles.
The prime meridian
was an arbitrary choice.
Columbus thought the world
looked like pear
and he was right. He had
no reason for this speculation.
Some facts are worth explaining.
Right now I can think only
of the wind in Chicago
and how I would rather
do anything with my life
than hurt yours.
There's more to making a map

than taking a few aerial photographs.
Invention isn't enough
and on this sea green planet
there's more going on than you know,
more than you can imagine.
How good a map is depends
on the ability of the mapper
to interpret what she sees on the surface.
I'm having a harder time with your face
than I had with the moon.
When continents bump up against
each other they leave scars and ridges
on the earth's crust. There's
a slow bleeding of lava
onto the ocean floor.
It took centuries to create
the delicate instruments of measurement
used by cartographers. Still,
triangulation from a known point
remains the basis of geodesy.

CHANGING WEATHER AT THE POINT OF DEPARTURE

Something important is not in the records
but everywhere are sinister clues
regarding recent and remarkable events in our town.
For some obscurity the table turned
and you were marching in with the parade
on the Fourth of July. Evidence,
impoverished and usually misleading
yielded to an ease of dinners on summer lawns.
Your life, clenched like a round spoon in the hand,
became distinct and important, worthy of protection.
There was a moment of entrance, the band leading
the way. Here was amazing reconciliation
with an old promise you had worked to forget;
the warm rain, the quiet air in the morning.
The rare geography of the moment caught you up.
You caught your breath.
The black trunk of the ancient tree splits
into two branches at eye level.
Thus Lopo Goncalves crossed
the equator. No boiling waters. No harm.

SENSIBLE SHOE IN THE GARDEN

At the edge of the well both hands are afraid:
   stone in pocket, green water still and waiting.
Great changes will occur.
Darwin was so sure.
I collect material for the study of variation:
   cryptic coloring, warning patterns, epigamic shades.
This is a code of sorts
   open to translation.
I am particularly interested in the transmission of design,
the long successful history of bony fish, unbroken
   by natural disaster.
      The sea green clear and blue clear.
Dandelions on the lawn, yellow and grey.
Some evolutionary moment caught me up,
a coincidence of persistent childhood habit
with larger botanical aims.
      I was making wishes.
The green grass clear and wet. Points of light
as points of departure.
      The airplanes take off:
the sky blue clear and cloudless.
It takes two hours to cross the ocean,
   small specks in the wind rounding
   out of sight.
Once they thought the world was a flat plate
surrounded by water.
      Blue clear and green clear.
The steady eye of the long gaze.
      You could weep real tears.
The sun shining into a well at noon changes everything.

REFUGEE

A PLOT

Imagine a yellow circle.
<u>Now</u>
bring me a flower;
resist me at every turn.

When you say certain things
I will write them down.
That way we know forever
that you have said them. Now.
I will begin now.

NIGHTWATCH

The rain in the city is falling somewhere else now,
over the ocean, moving into Africa.
I am tracing clouds of vertical development,
the swift progress of this latest storm.
Without these shadows we would change our shoes
the way a small boat struggles across the water.
We would get the words out.
Much there is that must be protected. Animals and birds
get nervous in such moments, such warmth
erupting out of season. Look.
Facts are nothing less than the courage
to name them, and we are, fortunately,
assisted in all difficult things
by nouns and verbs. A straight line
is the shortest distance
between two points.
Heart to brain, I would do it, I would risk it all
to see the sky cracked open with light.
If I could find where the lines are drawn
I would make you a map of this moment
turning into the next. I would give you
the illuminations of the night geographer,
the perfect dreams of this sudden drowning,
your whole life changing for the better.

SEX AND DEATH ON SULLIVAN STREET

I learn slowly
the way the just mind
splits in two   can say
yes and no   can say
I want scotch in a
glass   can say please
stay I want to talk
all night   can say please
go   please go home   you
misunderstand

Between us
where the space was
how many cups of coffee
I can't count   the language
suggests a body
and there is none; there,
we should like to say,
is a spirit.

A white night. Snow.
No more navigators.   Only
the map of anomaly to go on.
Signs occur eventually.
"Plage," to the blue water.
I read as "plague." A mistake.
Forgive my French, I mis-
perceived. We are happy
to take this train.
We love the jolly conductor
his blue jacket   his naive
approach to understanding.

DILEMMA

To
know what precedes and what follows
will assist yr/ comprehension of process

—Pound, Canto LXXVII

Now in that other place,
far from the land of parrots,
excuses are harder to come by.
What you were up to was tiring
though hardly necessary, a lot
like scrubbing floors on Saturday.
Anyway the sun was getting too hot.
It's enough to give yourself
a hard time about the wishbone,
the cricket sounds, the false starts
that keep you tense long after
the company has left for the evening.
You carry pictures around in your head,
ducks, decoys, the simpler
evasions of that old life.
The steady alternation of household tasks
contains a kind of felicity.
Every map has a point of view.
You write down the pictures in your head
and try to remember them exactly
as you saw them the first time.
Your breath aches when you run too far.
You run in the prairie
remembering that it burns each year.
The rabbit sits, the sun glowing
through its ears. It runs away.
You walk in the woods
picking wildflowers, knowing

only the name columbine.
This much gives you a certain
degree of authority over the situation.
You try to remember
the pictures of things the way they were
before the events happened.
Pretty soon you have trouble
remembering any of the events at all.
The sparrow sat and then flew away.
Then lots of things happen
and you try to sort them out.
This presents a challenge
and keeps you occupied
long past your bedtime.

THE FIRST SUBJECT MATTER FOR PAINTING WAS ANIMAL

In a cool spring I came to solve a mystery.
I had only one clear intention, the desire to study
particularities of temperature and light. At dusk
wild animals crept in from over the horizon. They belong
here and there, bringing messages, making promises. Mortal,
immortal, the dog brings fire instead of a bone.
Need is a mystery and I need you. Stock still, we were
animals caught in a moment of recognition, or an abyss
of miscommunication. I cannot say which.
This bread I bring you is the food of real meals.
We are free to leave at any time. There are at least
a million species of animals. Each may have many
ceremonies. The ritual of begging is only one possibility.
Assimilation is another. In the dark I could not hear
the cicadas but their sound was everywhere.

REFUGEE

I have a trigger for you, a bus stop, a way out.
I have a gift for you, a list in my hand.
The weather is always accurate because the mind just moves
    that way. Waiting
in the station I admire the multiple varieties of provincial
    life.
I'm red boots in the rain, a muffled alarm.

I make a living, I get along. I write tickets all the time and
    turn them in for things.

I've got an oven, a toaster, a house for birds.

You're always hungry, yes I can make soup, yes we can eat.
The kitchen tilts. Your dumb fists shift in their pockets.
You've turned on a dime, proud of your skill.
I'm mud at the edge of water.
The wind is moving fast. Yes, we can take a walk.

You have no scarf, no gloves, no excuse.
My shoes are crooked with salt; I cannot help you any more.
That one cries rags, that one sells hearts.
I'm telling you there's no end to it.
I have a map for you, a trigger, a sudden urge.

KEEPING TRACK

Distance to distance. The sound of
thunder, a train, recedes to the west.
Your hand running the length of my arm.
There is no beginning to an action.
You touch me, the sheet, your own
shoulders. We make love, seamless,
successful as the illusion of cinema.
"We have always been here," you tell me.
There we were. I am in a cafe
in New York in early spring.
"Coffee please." Afternoon to evening.
You are buying film in Columbus.
Distance to distance. I call
person to person. "Where does she sleep?
Will she be warm?" I was concerned.
"Here or there," you say, hugging
the phone or begging the question.

In San Antonio there were showers
of butterflies, monarchs bursting
the blue air with light. When I tell you
you remember the sun on the mountains,
exactly that color. I was surrounded
by butterflies and you were alone
on a mountain.

I lie down in a dark room.
Distance is nothing but time spelled out.
We were a film. We walked everywhere
in your camera. You are coming up
the stairs. I am waiting for you.
In bed we smoke cigarets with the

simple dignity of lovers about to be
serious. The smoke curls. It is
spring, summer, winter, fall.

The sheets are blue. They are clean.
We do not enter the bedroom. The
camera was there, is here. We sit
at a table. We are smoking cigarets.
My mind is light as a naked body.
We are exploding the blue air
with our bright thoughts. Only
the sequence suggests change. We
shoot it again. You are mounting
the stairs. I am counting
the necessary objects. The kitchen
table won't fit the frame.
The light is different in the evening.
I am waiting for a knock at the door.
I let you in. The table in the room.
The light in the evening.

Imagine a lie that unfolds
like a fan or a scream. The
objects are the same. The scene
unfolds like a fan and a scream.
One night, dreaming I could
no longer touch you, I awoke,
stayed silent, shivered until
morning. I fold you in my arms.
I unfold you. I am the shutter's eye.

You brush back the hair
from her face. She lies beside
you on the blue sheets in that
other room. I am a tree in
your yard. You are the wind
absent in the hour before the

hurricane. Everything has
already happened. There will be
a storm. What will she do when you
go into the yard to look at the tree?
There is the silence after thunder.
There is the wait for the rain.

Here has a name. There was
another place. Your name is
Michael. My name is Ruth.
Distance is distance. Ohio
is not New York. New York is
not California. California
is not Texas. A wooden house
is not a cafe on Third Avenue.
Once we met, we loved each other.
You ran your hand along the blue
vein in my arm. We made love
in the basement of a blue house.
Often there were stairs. There was
a door I opened many times.
There were trees with shadows
and wine in clear glasses. You
were from California. I was
from New York. There was
May and June and July.
There was a night and a
morning and a long afternoon.
We will die someday. The
camera needs film. Here
is a beginning. There is
a meal that ends, a storm
that passes; here is a tree
that sheds its leaves.

END OF CENTURY

END OF CENTURY—THE LIFE OF THE MIND

At last unable to do the simplest things
we finally took ourselves seriously.
But even that joke wore thin
after a while. All the time
we had been hoping for something different
other things were going on.

Like last night, for example,
going home, the express turned local *binary*
and suddenly we were stopping at all
these stations we'd never heard of.
At each one we waited and waited
but no one got on. Or off. When we
got home we slept like seeds and woke
in the morning. Breakfast: oranges.
Then you put on a new green jacket
and out you went while I waved
bye bye at the window.

This is always the starting point.
We take some unexpected journey
through the mountains and end up
where we were supposed to be all along.
It happens like this. And then. Then what?
Then watch out for the symbols, black
stones in the pockets.

For a while I carried
them everywhere, and for a lark, a song
I gave them away. I even gave some
to you, I remember, there by the
balustrade, in the old city. I gave them   *implying a new city*
all little names and gave them
away. For a lark. For nothing.   *key*

37

Now I am tired of this,
the way we are living. These
times. You know what I mean.
When was the last time you
had an easy night of it?
Things get bad for a while and then
they change to something else.
Maybe better. Maybe just different.

for better or for worse [handwritten]

not better or worse—just different [handwritten]

Once we were determined
to do it up right. A picnic.
A wild night in the morning.
How beautiful in your automobile.
Up into the hills with the top
down. Now I do the little things
I can count on: 1) make soup
2) mourn the dead 3) wash
the dishes 4) polish my nails 5) eat
an apple 6) enter the garden
7) wait wait wait
    stop
we know how this one
turns out

And then
the light in every room
more unexpected than sudden hunger.

—joy [handwritten]

Blooming trees every-
where, dripping seeds like snow,
weeks of sullen rain and oh
my heart is laden. Things
take their turn, slide
downhill into the river
facts run off the sides
of mountains.

I swear things will
never be the same
things never have
been you remind me.

remind herself—this is marriage [handwritten]

He loves me he loves me
not. In consequence I
adopted a certain limited
view of the situation
and clung to it hard.

likely *does this*
need to be
a consequence?

    draw me a map stupid
    we can't keep track — *she needs to know where things are &*
    of the story   *go.*

One potato two potato three
potato four five potato six potato
seven potato nor

on the seventh day of creation
he went wild.

    ok. stop.
    let's just stop this.

What can I tell you?
I was trying my best
to find an excuse
for both of us. I can't.
The accidents
of our making are everywhere.

There it is. The future.
Look at it, the run off, — *has chemicals*
smooth stones in the water,
this gift of ice and snow.
Enough water in every well.
Enough for the soup. Yes,
and for the docile cattle
by the pond under the trees
in the noonday sun. Remember
this, the simultaneous clinking
of glasses in the last few
sexual outposts of civilization.
Asks a lot to ask a whole
lot more than this.

But hunger there is
on the other side
of the map, stupid,
because the world is round
and not flat and always
was no matter where
you stand on the issue.

Chew down the drumstick.
Chew down the wishbone.
Then snap. Listen
you haven't got a chance
pal

Rumor has it
everything has already happened.
The architectural path is there,
leads back, ancient as Rome,
through layers and layers where
every situation is made strong,
at least in memory.

In those seven hills of memory
many things happened. Remember
how we were, then, late into the night
through need and time and the weight
of it and the desire to record it all
no matter how bad it got.

But let's give it a little credit
for surprising us once in a while.
We didn't know the season
was biased against us, that we'd worn
the wrong clothes out in the rain. That
same old rain. Relentless, evenhanded,
everyone, wise or not, getting
a piece of the action.

Now the story is almost over.
At the end of the day

you come home from the city
clutching a bunch of yellow
tulips, wearing the green jacket,
up from that hole in the ground,
up for a good dinner, soup and
bread and fruit, a little chat
about the old days. Next
we'll call the tailor and the
seamstress. We'll get the suit
and the white dress to fit the box
just so. We'll play it all out
till we know that at the inevitable
moment of arrival it will all be
so much less than we expected.

FATIGUES AND DISCOVERIES

THE DEFENESTRATION OF CHINA

Is this room a mistake?
Are wasps battering the windows?
Is a day darkening in late summer?

One can want to speak without speaking.
One could be a parrot, a gramophone

An infinitely long row of trees
is simply one that does not
come to an end.

, the idea of history

neverending stream of consciousness.

immortality of men w/ a unified consciousness.

IN THE COUNTRY

in memoriam: V. P. D.

And that and that and that and that and
that too. And then.

You lose your speech, your sight, you lose
your shuddering dreams.

You leave your little luggage on the stairs.
Out you go.

Then the water
runs down the windows.

After that
ordinary human suffering.

Now I can go anywhere.
I am leaving the room because you tell me to.

TERROR

you can't say it   you can't say
the color red is torn up
is bleeding   the color red
is pounded to bits

no bees in this garden   not
this year   no flight
of butterfly   only this
a wasp stings my ear

we say quiet
we say the pain lasts
only a little while
we say the red is vanishing

it hums in the blood
we clutch the idea
red we say   we bring it back
in the mind's eye

it starts as a fact
this little porch
this silence   this cry
at the heart's core

black hole, perimeter
immovable point at the center
say red say nothing
say nothing red nothing any more

THE SCENE OF THE CRIME

> "If we introduce the concept of knowing into the investigation it will be of no help."
>
> Wittgenstein, *Remarks on Color*

key for the whole book

I want some other set of instructions: I want
a satisfactory answer. I have this way
of knowing too much without knowing anything.
It comes upon me suddenly, like a stupid wind
from the west, like some lunar surprise.
It sits in my bones, it clamors for supper.
But then when I least expect it, on my way
to anywhere, there appears a small pool
of blood in the tunnel. Someone was cut
or shot or fell or was left bleeding
from the mouth. I am left at the periphery
of the disaster sifting stones one after
the other black to black white to white.
No light shines here. Comes the detective
with his holster and his gun and his
little silver pen. Dissolution
of voice, despair, then recovery. He knows
my story and tells it well. I grow older,
wise with random evidence, with incomplete
files. I repeat myself. I make myself clear:
I did not see I do not know I was not
there I did not hear I do not know
I do not know. All this attention
is too much of a good thing, a strong
cup of coffee handed over without
comment and without demand. I am
the bewildered one, the tired one.
I am the hungry one. I am the insecure
river hugging the shore. A coastline

blown up twice this size would have
the same unerring geometry. In slow
motion the plot does not thicken,
the soup does not boil.

                          A thin line
of blood leads from the mouth of the river
to a larger set of questions, to a sky
filled with tiny mirrors to sun, clouds,
blue snow. I will go, I say. I confuse
everything with the starlit sky,
with the off chance, with a fine
transparent stone:

Door; gate; bridge: let me go.

ARCTIC

"More than cold they hated the darkness"
—Barry Lopez, *Arctic Dreams*

The purity of intention remains
unresolved, no more evident to the eye
than the geographic north pole.
Take is comfort, then, in knowing this,
that the pole of inaccessibility
has been sighted several times
from the air. Aquatic animals,
birds that fly, think in three
dimensions. But when I stray
from the earth I am lost.
Each time I fall I learn
from the inside, bone in bone,
the hard crawl.
      The universe
is oddly hinged, a swinging door,
a thought surrounded by a halo.
I make up the rules as I go
along. Sometimes I alter them.
It is not always a question of choosing
the most complicated version.

STILL LIFE WITHOUT SUBJECT

Say it begins somewhere.
Say it is a journey.
Say you take it
while a glass of wine
gathers dust on a table.
Say that you once could tell
water from a flood of light
or those dark shapes
crowding the horizon.
Say you leave a dinner
framed by artichoke and pear.
Say you're waiting for a plane
or a photograph to be taken.

FERTILITY

the spider
used for prophecy
is an animal of wisdom

in its most
abstracted
form

the spider
is indistinguishable
from the frog

THE SHADOW OF A BIRD IN PASSAGE

The body is just a beginning, but I will go there,
to a lake in the western continent. Prepared for
travel, there is a time when earnestness and pure
love are necessary. I tell you you were born all
right. This is the beginning of a long story called
the body. It is difficult to be rid of the remnants
of evil propensities. Imagine horse eagles; recollect
revulsion. Do not go there by any means.

I am waiting at the edge of the water for a small
boat. We are face to face with great things, pumpkin bearing frogs, tarantula potatoes. It is difficult now to recall the dream of evil propensities.
A building is an act. Lately I have dreamed a lintel
and an arch. An act is a beginning. This is a time
when earnestness and pure love are necessary.
Recollect revulsion; do not go there by any means.

This is only the beginning of a long story. We are
face to face with great things, the body of a dream,
the soul choosing its moment of birth. There comes
a time in our lives when we burst our bonds or fail
to burst them. There is the dream of evil propensities. I tell you you were born all right. Beware
the broken facts of definition. Do not go there by
any means.

There are horses on the shore of the lake. They can
fly. The pure products of America go crazy; all thought
is in the present tense. We are face to face with
great things. The rest is talk. Pumpkin bearing frogs
sing in the dark: *I tell you you were born all right*.
The boat is steady in the water. Do not go there by
any means.

There is always the danger of sudden meetings. It is difficult now to recall the dream of evil propensities. There are four kinds of birth. The body is only a beginning. This is a time when earnestness and pure love are necessary. I have dreamed and dreamed the coming release of the heart. The body is an arch, a long sweet story. A sparrow falls, then flies again. A sparrow, in the form of a whale, picking crumbs from the street.

And in a bowl of water, one white peony, floating.

And Ptolemy choosing the Fortunate Islands.

Something about silver.

Before setting the table for dinner.

Any event about to happen.

A silver bridge bending in a glass of water.

The flower plucked from the garden.

Saying nothing, spending a long time at it.

The projective imagination interacting with topographical facts.

This geometry of expectation.

A sudden and natural generosity in the landscape.

Beached.

A single pear-shaped egg on the rock.

A symmetry of event.

Every map representing a particular problem.

Taking the whole situation into account.

You were thirsty.

The talk of others, a waterfall.

The cool glass in your hand.

The light bending to a point of concentration.

Sudden and late breaking events.

And freely, could drink.

Everything not said emanating urgent possibility.